I0478406

Data Integrity Pocket Guide

ISBN-13: 978-1546766179

ISBN-10: 1546766170

Contents

Introduction

In recent years, notified bodies and regulatory authorities such as the FDA have increasingly observed CGMP violations in relation to data integrity during CGMP inspections. The requirement for Data integrity is not new and is mandatory across both medical device manufacturers and producers of medicinal products. Data integrity is a key component of industry's commitment to ensuring the safe, effective and quality products reach the patient or end user.

These data integrity-related CGMP violations have led to numerous regulatory actions, including warning letters, import alerts, and consent decrees. The underlying premise in §§ 210.1 and 212.2 is that CGMP sets forth minimum requirements to assure that drugs meet the standards of the Federal Food, Drug, and Cosmetic Act regarding safety, identity, strength, quality, and purity.

Requirements with respect to data integrity in parts 21 CFR Part 211 and 212 include:

- ➤ Requiring that "backup data are exact and complete," and "secure from alteration, inadvertent erasures, or loss

- ➤ Requiring that data be "stored to prevent deterioration or loss"

- ➤ Requiring that certain activities be "documented at the of performance" and that laboratory controls be "scientifically sound"

- ➤ Requiring that records be retained as "original records," "true copies," or other "accurate reproductions of the original records"

➢ Requiring "complete information," "complete data derived from all tests," "complete record of all data," and "complete records of all tests performed"

Electronic signature and record-keeping requirements are laid out in 21 CFR part 11 and apply to certain records subject to records requirements set forth in Agency regulations, including parts 210, 211, and 212.

What does "data integrity" mean?

<u>FDA</u>

The FDA states that, data integrity refers to the completeness, consistency, and accuracy of data. Complete, consistent, and accurate data should be attributable, legible, contemporaneously recorded, original or a true copy, and accurate (ALCOA).

<u>MHRA</u>

The degree to which data are complete, consistent, accurate, trustworthy and reliable and that these characteristics of the data are maintained throughout the data lifecycle....such that they are attributable, legible, contemporaneously recorded, original or a true copy and accurate (ALCOA).

<u>PIC/S</u>

The extent to which all data are complete, consistent and accurate, throughout the data lifecycle.

What is "metadata"?

Metadata is the contextual information required to understand data. A data value is by itself meaningless without additional information about the data.

Metadata is often described as data about data. Metadata is structured information that describes, explains, or otherwise makes it easier to retrieve, use, or manage data. Among other things, metadata for a particular piece of data could include a date/time stamp for when the data were acquired, a user ID of the person who conducted the test or analysis that generated the data, the instrument ID used to acquire the data, audit trails, etc.

Data should be maintained throughout the record's retention period with all associated metadata required to reconstruct the CGMP activity The relationships between data and their metadata should be preserved in a secure and traceable manner.

What is an "audit trail"?

The FDA states that, "an audit trail means a secure, computer-generated, time-stamped electronic record that allows for reconstruction of the course of events relating to the creation, modification, or deletion of an electronic record."

An audit trail is a chronology of the "who, what, when, and why" of a record. For example, the audit trail for measurement system include the user name, date/time of the run, the integration parameters used, and details of a reprocessing, if any, including change justification for the reprocessing.

Electronic audit trails document the creation, modification, or deletion of data (such as processing parameters, alarm limits and results) and those that track actions at the record or system level (such as attempts to access the system or rename or delete a file). CGMP-compliant record-keeping practices prevent data from being lost or obscured.

Data Governance

The term "data governance" is seen becoming more familiar within life sciences. It encompasses the different elements that work to achieve data compliance and data integrity. (E.g. organizational controls, technical controls etc.) Data governance must be reviewed on an on-going basis to ensure the procedures are up-to-date and are been followed correctly.

Data governance can be divided into two categories, (1) Organisational and (2) Technical

Organisational controls include procedures and instructions for generation, completion and retention of data. Training is also a core element to ensure all staff are conscious of the importance of data integrity. The effectiveness of organizational controls must be verified with regular audits, surveillance and inspection of data records and practices. Technical controls depend largely on automation and Computerised Systems.

Design of Electronic Data Systems

Fundamental requirements of any electronic data system is to ensure the original data cannot be deleted. If there is a requirement for data to be manipulated, audit trails must be in place to show the type of change along with a reason, time, date, and identification of the person making the change.

Design of Paper Based Systems

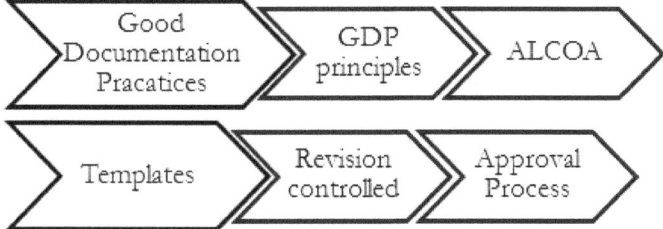

Traditional paper based systems for generating and managing CGMP records are still very much a part of quality systems within industry. While information technology has continues to transform the work envirnment (E.g. electronic master batch records), hardcopy, paper based systems are still common.

Paper generated records may be created by automated systems or by personnel. Where personnel are required to complete documentation by hand, procedures must be available in order to clearly set out the requirements and manner in which documents and completed and/or generated. Data Integrity of paper based systems is supported by the application of Good Documentation Practices (GDP) and the issue and control of approved templates and forms.

A Risk Based Approach to Data

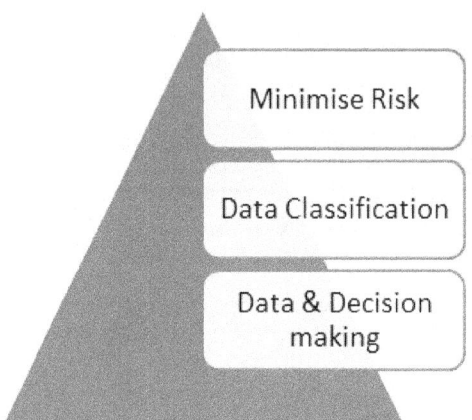

For Data Integrity to be best served, a risk based approach delivered through a Quality Management system. The above diagram highlights 3 steps to achieving effective Data governance.

Minimise Risk: reduce the potential risk to data integrity by training and the introduction of suitable controls

Data classification: Data may not impact patient safety. Some types of data is more critical than others. For example data in respect of cosmetic appearance of secondary packaging is not likely to cause injury to a patient. The defect may not be acceptable but less critical than other product or process related data which may impact decision making or impact upon safety or efficacy. Data should be classified based on the risk to patient or its role in decision making.

<u>Data criticality:</u> What is potential impact of the decision the data is intended to make? What decision does the data influence?

Key Terms – Electronic Data Systems

Configuration Identification

Software and hardware packages should be identified by a unique product identifier and a version number. For the software end-user, the parts of an automated system that are subject to configuration management should be clearly identified. The system should therefore be broken down into configuration items. These should be identified at an early phase of development so that a complete list of configuration items is defined and maintained. The application-specific items should have a unique name or version ID. The depth of detail when specifying the elements is decided by the needs of the system, and the organization developing that system.

Requirements for the User ID and Password

User ID: The user ID of a system should have a minimum length agreed with the customer and should be unique within the system.

Password: A password should always consist of a combination of numeric and alphanumeric characters. When setting up passwords, the number of characters and a period after which a password expires should be stipulated. The structure of the password is normally selected to suit the specific customer. The configuration is described in the section Security Settings of Password Policy. Criteria for the structure of a password are as follows:

> ➤ Minimum length of the password

> ➤ Use of numeric and alphanumeric characters
> ➤ Case sensitivity

Audit Trail

The audit trail is a control mechanism of a system that allows all data entered or modified to be traced back to the original data. A reliable and secure audit trail is particularly important in conjunction with the creation, change or deletion of GMP relevant electronic records. In this case, the audit trail must archive and document all the changes or actions made along with the date and time. Typical contents of an audit trail must be recorded and describe the procedures "who changed what and when" (old value/new value).

Data

Any data (numerical or otherwise) which is collected or processed as part of GxP activities in order to generate GxP documents and records using a paper-based or electronic process.

Data handling

Any GxP task that involves creation, entry, review, approval, analysis, reporting, storage, archival, retrieval, or disposal of GxP data

Data integrity

Degree to which a collection of GxP data is managed through effective organizational, operational, and technical mechanisms to ensure GxP data reliability.

Data lifecycle

The life-**cycle** starts from the time of data creation to the point of use and during its retention, archival, retrieval, and eventual disposal

GxP impacting

Any action that can impact the quality or safety of a product or critical process.

Application

Software installed on a defined platform/hardware providing specific functionality.

Bespoke/Customized computerised system

A computerised system individually designed to suit a specific business process.

Commercial of the shelf software

Software commercially available, whose fitness for use is demonstrated by a broad spectrum of users.

IT Infrastructure

The hardware and software such as networking software and operation systems, which makes it possible for the application to function.

Life cycle

All phases in the life of the system from initial requirements until retirement including design, specification, programming, testing, installation, operation, and maintenance.

Process owner

The person responsible for the business process.

System owner

The person responsible for the availability, and maintenance of a computerised system and for the security of the data residing on that system.

Third Party

Parties not directly managed by the holder of the manufacturing and/or import authorisation.

Regulations that speak to GxP and Data Integrity can apply to many different streams with the life science sector as previously mentioned. From medical devices to pharmaceuticals, all act in different manners, with long and short term applications. Take the example of a Total Knee Replacement. Many designs now ensure their effectiveness in excess of 10 years, even up to 20 years depending on individual circumstances. This requires many key records within manufacturing to be kept for several decades. Thus, data retention requirements specify the retention periods of such documents. The Integrity of GxP data must be protected during the entire data lifecycle. From creation of the data and records to the eventual destruction of data after the retention period is fulfilled.

Data integrity does not only apply to products it also applies to:

➢ Equipment
➢ Computerised Systems
➢ Test records
➢ Inspection records
➢ Material certificates

Data integrity ensures patient safety, product quality, and product supplies are generated by the product lifecycle processes. As such, the opportunities

Process Design

Failure to maintain data integrity can occur throughout the lifecycle of data; however, a thoughtful design of systems can prevent breaches in data and restrict the severity of any attempts to alter data. Therefore design, should aim to include controls and preventative measures. At a high level, this can be achieved by:

➤ Limiting access to GxP events and data
➤ Standard Operating Procedures (SOPs)
➤ Training
➤ System Owners

Data Reliability

Data reliability is the foundation to achieving cGxP data integrity. The FDA's ALOCA model can be used to enforce data reliability.

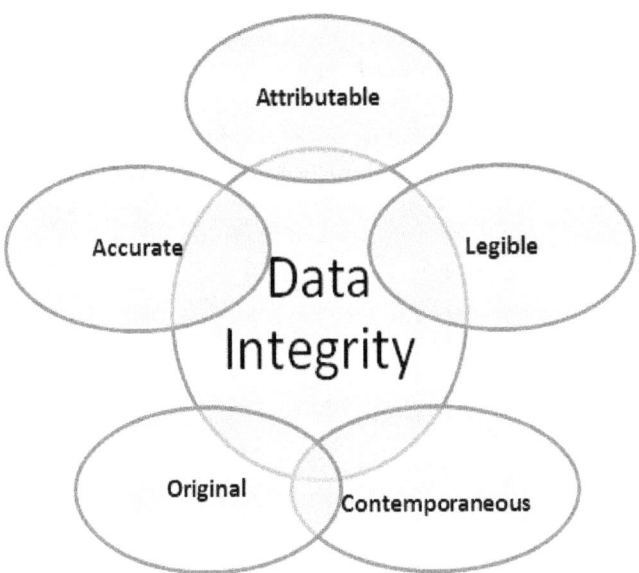

ALOCA graphical representation

Accuracy: the GxP data is recorded, calculated, analysed, and reported as found and correctly.

Attributable: any actions or calculations performance on GxP data can be attributed to or traceable to the person that performed the actions and the date and time at which they were performed.

Legible: the GxP data is recorded in a clear and human readable form

Contemporaneity: the GxP data is recorded at the same time as the observation/measurement is made or as soon as possible after the event.

Original: the initial data recorded is available and not altered. An additional point to make it that of trustworthiness. It is assumed that engineers and scientists etc. working across the Life science industries are ethical and do not falsify data or information. Typically companies can implement a code or practice or Ethical behaviour program to desist people from intentional unethical behaviour or the falsification of records.

The Lifecycle of Data

Data Creation: the point at which the values or data is created. The data and information is original (raw).

Data Authentication: Within a GxP environment, authentication refers to the approval of data (electronic signatures).E-signatures are key controls within software that prompt the user to enter a unique username and password to acknowledge a recording or action. The E-signature should create a permanent link with the electronic record that cannot be removed and can be viewed through an audit trial.

Data Protection: Once the data is created, the handling of the data must ensure data integrity. For electronic data, this includes access control to computer systems. Other practical restrictions can also be made such as limiting room and site access to authorised personnel.

Data Retention: The controlled storage, backup and arching of data. Retention of records may be required for several decades depending on the type of data and the regulatory requirements relating the particular product or industry.

Technical Controls

The benefits of modern software and computerised systems allow robust and complex data handling and calculations to be completed. With this modern capability that is becoming more powerful, comes more responsibility with regard to the use of data.

The Computerised systems used to generate, gather or interpret GxP data must be fulfil several criteria. First and foremost they must be fit for the intended use. The software and hardware must be validated and proven to be consistent and reliable. Some general considerations for the use of Computerised systems include:

➤ Design to foster integrity of GxP data
➤ User requirements specification detailing the intended use and required functionality
➤ An approved vendor with certification to ISO 9001 or other Quality management standards
➤ Software should meet the requirements of regulations such as FDA 21 CFR Part 11.
➤ Written procedures on how automated processes function.

It should not be an easy process from persons to alter or corrupt data when using computerised systems. GxP impacting Computer systems should have controls that prevent unauthorised access along with audit trail history.

Audit trail design and configuration capture key critical processes, events, settings and information. This enables any investigations of quality events impacting data integrity to be reviewed and analysed.

Organisational Controls

Regulated companies such as medical device, pharmaceutical and biotechnology companies are required to operate under a Quality Management system. For medical devices, ISO 13485 serves as a Quality Management System. Likewise, the FDA Code of Federal Regulations 21 CFR Part 211 for finished pharmaceuticals.

Organisational controls for Data Integrity can address:

- ➢ Assessment of GxP computerised systems
- ➢ Management of GxP computerised systems
- ➢ Electronic Records Implementation and handling
- ➢ Use of Electronic signatures
- ➢ Quality Risk Management

Operational Factors

Operational factors refer to process or manufacturing errors, deviations or non-compliance to established procedures that may impact data integrity.

GxP data handling activities should be designed to limit human intervention. As with human intervention there can errors or omissions. Furthermore, it may call in to question the reliability of the data.

Mistaking proofing methodologies should be developed to avoid human error related breaches in data integrity. As with any system or technology training is a fundamental step. Building upon training, exposure to GxP data systems and On the job training all play a part in delivering a system that is robust and meets regulatory requirements. It is important to remind ourselves that while regulations are the driving force to comply with Data integrity, it is for the protection and safety of the patient or end user of the product, medicine or treatment.

Practical Elements to Data Integrity

Facilities and systems must be configured in a way that encourages compliance with principles of data integrity. Examples include:

- ➤ Availability of clocks for recording times.
- ➤ Access points to allow swift reference to GxP records at locations where tasks are completed.
- ➤ Control of raw data.
- ➤ Control of approved documents.

Computer system design and development

For Computer systems, software requirements are typically stated in functional terms and are defined, refined, and updated during the development phase. Success in accurately and completely documenting software requirements is a crucial factor in successful validation of the resulting software. A specification* is defined as "a document that states requirements." It may refer to or include Engineering drawings or other relevant documents 21 CFR 820.3(y).

There are different kinds of written specifications:

- ➢ User requirements specifications
- ➢ System requirements specification
- ➢ Software requirements specification
- ➢ Software design specification
- ➢ Software test specification
- ➢ Functional Design Specification

All of these documents establish "specified requirements" and are design outputs for which various forms of Verification or Validation are required. The URS must also define non-software requirements and hardware. Non-functional requirements such as maintainability and usability can also be included. There should be a clear distinction between mandatory regulatory requirements and optional features. Proper definition at this stage ensures the system meets Data integrity requirements and prevents costly updates down the line.

Software Validation

Where there is the potential to affect product conformance to requirements or where software or IT systems provide support to aspects of Quality Management, validation is required.

Most companies categorise software validations to account for the different applications of software and IT systems. For example, Enterprise systems, such as the drawing package SolidWorks would be validated in a different manner to Manufacturing Systems that contain software (a.k.a. embedded software).

"Embedded" software is where the software is integrated into the manufacturing equipment. Embedded software is typically validated during the Equipment Qualification stage, Process Validation stage or Test Method Validation. Enterprise software falls outside of Equipment or Process Validation but does require validation if it impacts product quality or is used to make quality decisions. Standalone systems such as ERP (Enterprise Resource Planning) systems also require validation.

Good Automated Manufacturing Practice (GAMP) is a set of guidelines for manufacturers and users of automated systems in regulated industries. Specifically, the Medical device, pharmaceutical and biopharmaceutical industries. The application of GAMP and Validation of Automated Systems in manufacturing helps ensure that regulated medical devices and medicinal products have the required quality and are manufactured according to Good practices, meet regulatory and legal requirements and ensure patient safety. GAMP ensures quality is in-built into each stage of the manufacturing process. Therefore, GAMP has a place in all aspects of automation and production, including the handling of raw materials, control of facilities and equipment etc.

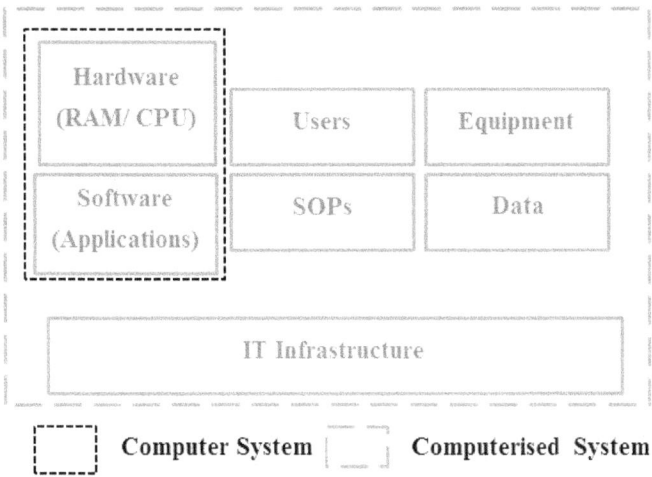

| Hardware (RAM/ CPU) | Users | Equipment |
| Software (Applications) | SOPs | Data |

IT Infrastructure

Computer System　　Computerised System

Automated System: Term used to cover a broad range of systems, including automated manufacturing equipment, control systems, automated laboratory systems manufacturing execution systems and computers running laboratory or manufacturing database systems. The automated system consists of the hardware, software and network components, together with the controlled functions and associated documentation. Automated systems are sometimes referred to as computerised systems; in this Guide the two terms are synonymous.

Commercial off-the-shelf (COTS): Configurable Programs- Stock programs that can be configured to specific user applications by "filling in the blanks", without (COTS) altering the basic program.

Computer System Validation: a process that confirms by examination and provision of objective evidence that the computer system conforms to user needs and intended uses. System validation is a process for achieving and maintaining compliance with GxP regulations and fitness for intended use by adoption of life cycle activities, deliverables, and controls.

GAMP 5: is a set of guidelines that offers a Risk-Based approach to ensuring the compliance of GxP impacting computerised systems.

V- Model: is a development process which sets out a roadmap of stages and deliverables during a project.
21 CFR Part 820: FDA requirements pertaining to Medical Devices.

User Requirement Specification, URS: The URS is a critical document that defines the requirements of the computerised system and agreement to the requirements.

Software Requirement Specification, SRS: an SRS can be written to interpret the requirements of a URS and how they relate to the requirement or how the requirement is met in practical terms regarding software.

Functional Design Specification, FDS: a functional design specification is a document that specifies how particular requirements are met – this can be a combination of how the equipment/process operates mechanically/automatically etc. An FDS is typically written to response to a URS.

Computer System Validation Life Cycle

The Computer System Validation Life Cycle refers to all activities from initial concept to retirement of a computer system. The life-cycle of the system includes the defining of, and performance of activities in a systematic way from conception, requirements, development or configuration, testing, release and operational use.

The four GAMP Life-cycle phases include:

➢ Concept

➢ Planning and Project stage

➢ Operation

➢ Retirement

The Concept Stage is concerned with understanding the need or the problem to be addressed. We will see that the User Requirement Specification (along with other specifications) and the initial risk assessment help to drive a project forward in a systematic manner. The most common life-cycle approach for Computerised and Automated systems is the V-Model. The GAMP based V-model lays out a roadmap which facilitates the Validation of equipment and automated systems.

The Planning and Project stage involves the planning of the validation effort required to implement the system into the business area(s) based on identification and approval of system concept. This phase includes assessments of the regulatory and system risks, supplier assessment, development of validation strategies, identification of deliverables that will be generated, definition of the business process the system will support as well as the user requirements which the system will fulfil.

Design & Development   and configuration of the hardware and software is also required to meet the system requirements as per specifications. In case of custom Software components, this effort could also include detailed Software design and developmental testing to ensure readiness for verification testing.

Verification – This effort confirms that specifications have been met and releases the system for use. This phase will involve multiple stages of reviews and testing depending on the system type, the development method applied and its use. Once verification activities have begun any changes to the system must be captured through change control.

On successful completion of the verification activities, the system is then released for effective use. The Test strategy and other verification activities will vary widely between simple equipment and more complex customised/configurable systems. The verification and validation approach is typically agreed and detailed in at the validation planning stage. The VP can be updated accordingly as the project develops with more detail been added. Alternatively, a test strategy document or matrix could be written to provide more specific test plans.

Verification deliverables vary based on the complexity and level or customisation of the system in question. Corporate or company specific procedures also shape the required activities to be completed and reported. Some generic deliverables are listed below.

- Approval, executing and review of test protocols
- Writing and approving SOPs for operation and maintenance of the system
- Traceability Matrix
- Completion of any Risk mitigations (e.g. updates to FMEA etc.)
- Validation Summary Report(s)

Validation reporting requirements varies depending upon the scope of the system and should also be driven by a procedure and template. The Validation Plan can also outline the deliverables and what needs to be addressed in the report. A Validation Summary Report (VSR) shall be written which summarizes the results of executing the VP the documents created for the validation activities, summarizes (or points to summaries) of the testing performed. Finally, the VSR indicate the acceptance of the system/equipment by the user by the Project team and state that the equipment is released for commercial operation / production.

The operation phase supports the need to maintain compliance and fitness for intended use after the system is released for normal use. It is important to ensure the system remains within a continued validated state. All proposed or necessary changes to the system must be assessed and controlled as part of a change control process. Once the system has been accepted and released for use, the operation phase begins. This phase consists of maintaining the system's compliant state and fitness for intended use through the control of the procedures supporting the system's operational use.

During the operation phase the below activities are typically completed:

- Ongoing Training
- Preventative Maintenance
- Service management and performance monitoring.
- Change Control
- Periodic review
- Maintaining system security
- Records management
- Calibration

The retirement phase involves the planning and proper management of activities relating to the removal of systems from service (shutdown). The retirement should take into account the storage of any data and any data migration that needs to occur prior to retirement. The retirement plan, if needed, will outline the retirement strategy from the roles and activities that will be conducted to the removal of the system for use. A Retirement Summary Report is produced that documents the results of the activities defined in the retirement plan including:

- Retirement Plan and Timelines.
- Summaries of any data migration activities.

- Identification of the storage location of documentation relating to the system.
- Obsoleting of SOPs.

It must be stressed that GAMP is a set of principles, a set of guidelines that aim to achieve compliant computerized systems that are fit for intended use. GAMP Guidelines differ to 21 CFR QSR regulations as they are not legal or statutory requirements. However, they represent industry best practice and compliment the Validation efforts that are legal requirements and statutory requirements.

Software Validation is a requirement of the Quality System regulation, 21 Code of Federal Regulations (CFR) Part 820. Validation requirements apply to:
(1) software used as components in medical devices,
(2) software that is itself a medical device, and
(3)software used in production of the device or in implementation of the device manufacturer's quality system.

Note: EU GMP Annex 11, provides information on the inspection of 'Computerised Systems'.

Examples of Security requirements:

- Three levels of access required, operator, and engineer and maintenance
- Engineer - access to all screens, to modify process settings Maintenance - access to functions required to perform machine maintenance activities.
- Operator - restricted access, does not have access to change process settings.
- Different access levels will require different passwords.
- No security will be required for basic operations (Start/Stop)
- A user auto-logoff feature shall be incorporated in the

design. The auto- logoff time shall be configurable.

- A soft copy of Program settings must be provided with delivery of the equipment.

Examples of Security requirements:

- The reject count and yield must be displayed on the HMI screen.
- Real-time readings for all critical parameters shall be visible on the HMI Screen.
- All Critical parameters shall be adjustable via the HMI Screen.
- The status of each door should be visible on the HMI screen.

System Categorisation

GAMP 5 makes provision for four categories of software in order to distinguish the level of customization/configurability that exists across software's serving different functions.

GAMP Software Category 1, Operating Systems
GAMP Software Category 2, Non-configured software
GAMP Software Category 4, Configurable software packages
GAMP Software Category 5, Custom Software

GAMP Software Category 1, Operating Systems

Category 1, operating systems, covers established commercially available operating systems.

These are not subject to validation themselves, the name and version of the operating system must, however, be documented and verified during Installation Qualification (IQ). Application software hosted on operating systems need to be validated.

GAMP Software Category 3, Non-configured software

Category 3 covers commercially available, standard software packages and "off the- shelf" solutions for certain processes. The configuration of the software packages should be limited to adaptation to the runtime environment (for example network and printer connections) and the configuration of the process parameters. The name and version of the standard software package should be documented and verified in an Installation Qualification (IQ). Special user requirements, such as security, alarms, messages, or algorithms must be documented and verified in an Operational Qualification (OQ).

GAMP Software Category 4, Configurable software packages

GAMP Software Category 4, Configurable Software Packages Category 4 covers configurable software packages that allow special business and manufacturing processes. This involves configuring predefined software modules. These software packages should only be considered as belonging to Category 4 if they are well-known and mature. Normally, a supplier audit is necessary. If this is not available, the software packages should be handled as Category 5. The name, version, and configuration should be documented and verified in an Installation Qualification (IQ). The functions of the software packages should be verified in terms of the user requirements in an Operational Qualification (OQ). The Validation Plan should take into account the lifecycle model and an assessment of suppliers and software packages.

GAMP Software Category 5, Custom Software

GAMP Software Category 5, Custom Software Custom/Bespoke Software (GAMP Software Cat 5) is software that contains custom code designed or modified specifically for a particular customer. As the code is custom it presents a greater risk. This risk must be mitigated with the right approach to the validation.

GAMP Considerations

Correctly assigning a GAMP software category to equipment, a system or process is an important activity that should be completed early-on in the planning stage of a project. There must of some degree of familiarity with the equipment or system. The manufacturer or vendor can be a source of information that may help the designation. In many cases, companies create tools or processes that help determine what GAMP software category applies. These have different names such as questionnaires, screening tools, planning tools etc.

Risk Assessments

A Risk Assessment process should be applied to cGxP computerized systems in order to identify and mitigate potential risks to (1) patient safety, (2) product quality and (3) data integrity. Results identified through a Risk Assessment help to determine the validation strategy, the effort and time required, and allow better targeting of the validation activities to the highest risks.

The Risk Assessment should be revised during the Software Development Lifecycle (SDLC) if the functionality, requirements or intended use of the system changes. The Risk Assessment activity should also be evaluated during system build-up as well as when implementing changes. Risk Assessment tools for cGxP computerized systems are typically completed during the planning stage, specification stage and post qualification if a change or update is required.

Planning Stage

Initial Impact/Risk Assessment – during the planning phase to identify the level of impact and GxP relevance of the system/equipment. (Tools used: High Level Risk Assessment).

Specification Stage

Functional or Quality Risk Assessment – during the specification phase - identify potential risks and possible mitigations to be to be introduced to the process. (Tools used: Quality Risk Matrix, (p)FMEA).

Changes to the system

Impact Assessment of changes – as part of the change control process in the system operational phase. The following diagram defines the Risk Assessment steps within the System Life Cycle (Tools used: Impact assessment checklist, Change control procedures).

Quality Risk Matrix

A QRM is a risk assessment that identifies and manages the risk to patient safety, product quality and data integrity that relate to the systems processes. Risk Scenarios or potential causes should be developed for each identified function or process step and then assessed for the impact on patient safety, product quality or data integrity. Risk mitigations and controls should then be introduced to address both medium and high levels of risk. The QRM requires 3 "assessments" in order to produce an estimation or overall Risk (Low, medium, high)

> ➢ Assess Likelihood
> ➢ Assess Detectability
> ➢ Assess Severity

Traceability Matrix

A Traceability Matrix should be prepared as required in accordance with company and internal policy. It is also recommended by GAMP guidelines, ASTM E2500 and ISPE Risk based approach to Validation. The matrix links the user requirements and specifications to the testing and validation activities. A traceability matrix illustrates that all user requirements are traceable to the verification/validation activity or vendor documents as relevant (FDS if applicable, Design specifications etc.) A simple traceability matrix (TM) format is shown below on the next slide. Generally, individual organisations will have an approved template to work from. However, the URS structure can form the basis of the template, with additional columns added to document the test/verification method, Reference documents (such as FDS' and vendor specifications and design documents)

21 CFR Part 11

This section specifically covers the regulatory requirements of part 11 of Title 21 of the Code of Federal Regulations; Electronic Records; Electronic Signatures (21 CFR Part 11).

Part 11 of the FDA CFR is relevant to "records in electronic form that are created, modified, maintained, archived, retrieved, or transmitted under any records requirements set forth in Agency regulations." This first section of the book provides a background information and explanations of each section and requirement of the regulation. The second half of this eBook provides a clear and transferrable verification process for each requirement of 21 CFR Part 11, with suggested verification methods included.

As of 2007, several sections of the regulation have been identified as excessive and the FDA announced in guidance that it will exercise enforcement discretion on some parts of 21 CFR part 11. This has been welcomed by some manufactures but it has also causes a degree of confusion.

The requirements relating to access controls are the most fundamental requirements and are routinely enforced. The "predicate rules" that required organizations to keep records the first place are still in effect. If electronic records are illegible, inaccessible, or corrupted, manufacturers are still subject to those requirements.

If a regulated firm keeps "hard copies" of all required records, those paper documents can be considered the authoritative document for regulatory purposes. This then means that the computer system is not in scope for electronic records requirements, although subject to predicate rules which still require validation.

If the "hard copy" is to be identified as the authoritative document, the "hard copy" must be a complete and accurate copy of the electronic source. The manufacturer must use the hard copy (rather than electronic versions stored in the system) of the records for regulated activities.

Definition of Records

The FDA has deemed the following records or signatures in electronic format subject to 21 CFR part 11:

"Records that are required to be maintained under predicate rule requirements and that are maintained in electronic format in place of paper format. On the other hand, records (and any associated signatures) that are not required to be retained under predicate rules, but that are nonetheless maintained in electronic format, are not part 11 records.

Records that are required to be maintained under predicate rules, that are maintained in electronic format in addition to paper format, and that are relied on to perform regulated activities. Records submitted to FDA, under predicate rules (even if such records are not specifically identified in Agency regulations) in electronic format (assuming the records have been identified in docket number 92S-0251 as the types of submissions the Agency accepts in electronic format). However, a record that is not itself submitted, but is used Contains Nonbinding Recommendations in generating a submission, is not a part 11 record unless it is otherwise required to be 205 maintained under a predicate rule and it is maintained in electronic format.

Electronic signatures that are intended to be the equivalent of handwritten signatures, initials, and other general signings required by predicate rules. Part 11 signatures include electronic signatures that are used, for example, to document the fact that certain events or actions occurred in accordance with the predicate rule (e.g. approved, reviewed, and verified)."

The above definitions are taken from FDA guidance document entitled "FDA Guidance for Industry: 21 CFR Part 11 - Electronic Records and Electronic Signatures:

Scope and Application, August 2003." This document also provides recommendations on documenting key decisions that may be taken in relation to 21 CFR Part 11 applicability and compliance.

Requirements and Specifications

The need for compliance to 21 CFR depends on type of technology and level of automation and computerisation involved in the manufacturing process or other actives that are GxP impacting. Does the system store electronic records? Does the system require a login? Is there an audit trial? If a complex system is to be procured, the requirements need to be communicated to the manufacturer as part of a User requirement specification and/or software requirement specification.

General Guidance on Requirement Specifications

While the Quality System regulation states that design input requirements must be documented, and that specified requirements must be verified, the regulation does not further clarify the distinction between the terms "requirement" and "specification." A requirement can be any need or expectation for a system or for its software. Requirements reflect the stated or implied needs of the customer, and may be market-based, contractual, or statutory, as well as an organization's internal requirements.

There can be many different kinds of requirements (e.g., design, functional, implementation, interface, performance, or physical requirements). Software requirements are typically derived from the system requirements for those aspects of system functionality that have been allocated to software. Software requirements are typically stated in functional terms and are defined, refined, and updated as a development project progresses. Success in accurately and completely documenting software requirements is a crucial factor in successful validation of the resulting software. A specification is defined as "a document that states requirements." (21 CFR 820.3(y)) It may refer to or include drawings, patterns, or other relevant documents and usually indicates the means and the criteria whereby conformity with the requirement can be checked.

There are many different kinds of written specifications, e.g., system requirements specification, software requirements specification, software design specification, software test specification, software integration specification, etc. All of these documents establish "specified requirements" and are design outputs for which various forms of verification are necessary.

Validation of Computerised Systems

The requirement for computerised systems to be compliant to 21 CFR part 11, needs to be identified early on the project to ensure that the vendor or supplier of the systems or equipment can develop, build a system that meets the requirements of 21 CFR part 11. Computer system validation can be divided into 3 distinct phases which include: (1) Plan, (2) Design & Development, (3) verification and (4) Retirement. The requirement for computerised systems to be compliant to 21 CFR part 11, needs to be identified early on the project to ensure that the vendor or supplier of the systems or equipment can develop, build a system that meets the requirements of 21 CFR part 11.

Plan: This phase involves the planning of the validation effort required to implement the system and identification of key milestones and requirements. It requires supplier assessments, assessments of the regulatory and system risks, supplier, development of a validation approach and the identification of deliverables that will be generated, that will support the implementation and operation of the system.

Design & Development: This phase consists of the design, development and configuration of the hardware and software required to meet the system requirements. In case of custom software, design and developmental testing is important to ensure proper functionality prior to verification testing.

Verification: This phase confirms that requirements and specifications have been met. Testing is required to ensure the system operates as intended. Upon successful testing and verification, the system can be released for use. Once verification activities have begun any changes to the system must managed through change control. In case of successful completion of the verification activities (i.e. any deviation has been evaluated and addressed), the system is released for effective use. Operation This phase supports the need to maintain compliance and fitness for intended use after the system is accepted and released for use.

Retirement: This phase consists of the planning, executing and summarizing of the events required for system shutdown. It includes the appropriate handling of the supporting documents and the data contained within the system. While described here as a separate phase, a system's retirement can be handled as part of a new system implementation or as a separate project.

Best practice when it comes to Computer System validation is to adopt a life cycle approach for computer systems which requires the completed of activities in a systematic way from system conception to retirement. Life cycle activities could be scaled according to system impact on product quality, patient safety and data integrity, system complexity and novelty, supplier assessment and business risk.

Computer System: A computer / automated system consisting of the hardware, software, and network components, together with the controlled functions (personnel, procedures, and equipment) and associated documentation.

Computer System Validation: A process that confirms by examination and provision of objective evidence that the computer system conforms to user needs and intended uses. Computer System validation is a process for achieving and maintaining compliance with GxP regulations and fitness for intended use by adoption of life cycle activities, deliverables, and controls.

GxP Regulated Computer Systems: Computer systems determined to have a potential impact on Product Quality, Patient Safety and Data Integrity; these systems are required to comply with the relevant GxP regulations.

Data Integrity: is the degree to which data is reliable and without error. Data must be accurate, attributable, contemporaneous, original, legible and available. A breach of data integrity occurs when any person manipulates or distorts data and submits the results of that data as valid.

Predicate rules: a predicate rule is any FDA regulation that requires companies to maintain certain records and submit information to the agency as part of compliance.

To gain a better understanding of the validation of computerized systems, consult the following publication- "FDA's guidance for industry and FDA staff General Principles of Software Validation." Industry guidance such as the GAMP 5 guide issued by ISPE is also a useful reference.

Electronic Records

When it comes to the regulated industries such as the medical device industry, every process and procedure must be documented. Documentation ensures that everyone is working in the same manner with the same procedures. However, documentation is more than just writing down procedures and processes. It is also concerned with how documents are controlled, how they are updated and how they are stored.

Electronic Document management systems

Electronic document management systems aka EDMS are now the norm and gold standard for most medium to large organisations. Many companies that provide medical device manufacturers with an EDMS can be customised to match the business processes particular to an organisation. With configurable or customisable software, validation and proper verification is important to ensure the system operates as intended. There are also regulatory requirements that stipulate the expectations and requirements of such system. For example, the application of electronic signatures and the presence of audit trials. FDA 21 CFR Part 11 details the requirements with regards to electronic records and electronic signatures. For medicinal products in Europe, GMP V4 Annex 11 specifies similar requirements.

Record Retention

Regard to the part 11 requirements for the protection of records to enable their accurate and ready retrieval throughout the records retention period (11.10 (c)) Persons must also comply with all applicable predicate rule requirements for record retention and availability such as (211.180(c) general requirements. The decision to follow 21 CFR part 11 should be justified and documented as part of a risk assessment and based on the value of the records over time.

FDA does not object to archiving of required records in electronic format to non-electronic media such as paper, or to a standard electronic file format (examples of such formats include, but are not limited to, PDF, XML, or SGML). Persons must still comply with all predicate rule requirements, and the records themselves and any copies of the required records should preserve their content and meaning. As long as predicate rule requirements are fully satisfied and the content and meaning of the records are preserved and archived, you can delete the electronic version of the records. In addition, paper and electronic record and signature components can co-exist as long as predicate rule requirements are met and the content and meaning of those records are preserved.

Electronic Signatures

Electronic signatures are computer-generated character strings that count as the legal equivalent of a handwritten signature. The regulations for the use of electronic signatures are set out in 21 CFR Part 11 of the FDA. Each electronic signature must be assigned uniquely to one person and must not be used by any other person. It must be possible to confirm to the authorities that an electronic signature represents the legal equivalent of a handwritten signature. Electronic signatures can be biometrically based or the system can be set up without biometric features.

Conventional Electronic Signatures

If electronic signatures are used that are not based on biometrics, they must be created so that persons executing signatures must identify themselves using at least two identifying components. This also applies in all cases in which a chip card replaces one of the two identification components. These identifying components, can, for example consist of a user identifier and a password. The identification components must be assigned uniquely and must only be used by the actual owner of the signature.

When owners of signatures want to use their electronic signatures, they must identify themselves by means of at least two identification components. The exception to this rule is when the owner executes several electronic signatures during one uninterrupted session. In this case, persons executing signatures need to identify themselves with both identification components only when applying the first signature. For the second and subsequent signatures, one unique identification component (password) is then adequate identification.

Change Management

Validation programs are subject to change control. Each company or organisation should have a procedure detailing the change management process. Below is a suggested overview of a typical change control process.

Any system, facility, document or process that has the potential to impact product quality and validated state is generally subject to following a change control process.

Another term used in industry is Enterprise Change Control or Engineering Change Control. Essentially these terms are the same. The intent is to control and manage change consistently.

A change control can take the form of a document which drives the agenda and the specific requirement. Change control is also created with enterprise software such as Kintana, Documentum and SAP. While each company will have varying processes, some basics are common. These include the 3 stages of change control; pre-implementation, implementation and post implementation (if required).

Validation Deliverables

The deliverables of validation activities should be in accordance with a Project Validation Plan of Validation Master Plan. For small projects or changes to Computerised systems, a change control may serve as the Validation Plan. However, some typical deliverables include the following:

- ➢ GxP Assessment (note, some systems may be non GxP applicable)
- ➢ User requirements specification
- ➢ Third party audit
- ➢ Validation plan
- ➢ Design specification such as functional, software, hardware and technical specifications
- ➢ GxP risk assessment
- ➢ Validation protocols
- ➢ Traceability matrix
- ➢ Validation report.

Summary of 21 CFR Part 11.10

11.10 (a) Accuracy, reliability & consistent intended performance.

11.10 (b) Copies of records (Paper)- complete copies of records in both human readable and electronic form suitable for inspection, review, and copying.

11.10 (b) Copies of records (Electronic)- complete copies of records in both human readable and electronic form suitable for inspection, review, and copying.

11.10 (c) Protection of records- accurate and allow ready-retrieval throughout the records retention period.

11.10 (d) & (g) Authorised access - Limiting system access to authorized individuals in relation to accessing records, the operation or computer system input or output devices, altering of records, or performing the operation at hand.

11.10 (f) Sequencing of steps - checks to enforce permitted sequencing of steps and events.

11.10 (h) Input device authorisation - checks the validity of the source of data.

11.10 (i) Input device persons education, training, and experience.

11.10 (j) Establishment of written policies - establishment of, and adherence to, written policies that hold individuals accountable and responsible for actions initiated under their electronic signatures, in order to deter record and signature falsification.

11.10 (k) Appropriate Controls - appropriate controls over systems documentation-(1) Adequate controls over the distribution of, access to, and use of documentation for system operation and maintenance.

(2) Revision and change control procedures to maintain an audit trail that documents time-sequenced development and modification of systems documentation.

Electronic Records Verification Methods

(21 CFR 11.10 Electronic Records) some simple verification methods for electronic records

11.10 (a) Accuracy, reliability & consistent intended performance

Verification Method: create test scripts to verify that all types of records generated and maintained by the system are accurate, consistent and contain the intended data. Repeat tests using different challenge conditions to cover any anticipated operating conditions. Ensure that test scripts contain the relevant acceptance criteria.

11.10 (b) Copies of Paper based records

Verification Method: Create paper copies for the record types under test and verification. Compare the paper hardcopies with electronic records that are stored and displayed in the computer system.

11.10 (b) Copies of electronic records

Verification Method: Create electronic copies (softcopies) and compare the copy to that contained in the computer system.

11.10 (c) Protection of records

Verification Method: Review the controls and procedures that ensure accuracy and ready retrieval throughout the records retention period.

11.10 (d) & (g) Authorised access

Verification Method: Complete security verification for each account type e.g. operator manager etc. Verify that access is granted and denied via the login function.

11.10 (f) Sequencing of steps

Verification Method: The purpose of this requirement is to ensure the correct sequencing of steps and events occurs. Create tests scripts to demonstrate that any series of steps are operating as intended e.g. approval or signature or login.

Note: This requirement relates to electronic records and is not intended to verify any Operational or functional sequences. These are typically covered in Equipment Validation.

11.10 (h) Input device authorization

Verification method: The purpose of this step is to check the validity of the source of data. When data is networked or transferred form a device the user must verify that the device is a valid input device. This can be achieved by code review or physical testing.

11.10 (i) Input device person's education, training, and experience

Verification method: Ensure persons using the system are appropriately educated, trained and experienced. Attach evidence of same.

11.10 (j) Establishment of written policies

Verification method: verify the availability of written policies in relation to responsibilities for actions taken under their electronic signatures.

11.10 (k) Appropriate Controls - appropriate controls over systems documentation

Verification method: review access to SOPs to verify their distribution and use for operation of the system is controlled. Verify that chance control is practiced for system related documentation.

What is a Quality Risk Matrix?

A QRM is a risk assessment that identifies and manages the risk to patient safety, product quality and data integrity that relate to the systems processes.

Risk Scenarios or *potential causes* should be developed for each identified function or process step and then assessed for the impact on patient safety, product quality or data integrity. Risk mitigations and controls should then be introduced to address both medium and high levels of risk.

<u>Quality Risk Matrix and 21 CFR Part 11</u>

To ensure proper compliance to 21 CFR Part 11, a simple Risk assessment can help to identify any gaps or concerns. Below an example of a Quality Risk Matrix approach is shown. As this example is a top level initial assessment, the risk classification is estimated as Broadly acceptable, intolerable or as low as possible. This determination is based on the information available and application of relevant experience of dealing with the vendor in question and validation of automated systems.

As a project moves through the phases of planning and design and development, a more detailed risk assessment can be created. At this point a more detailed estimation or risk classification can be made. This can be done by applying 3 "assessments" in order to produce an estimation or overall Risk (Low, medium, high). The 3 assessments are: (1) Assess Likelihood, (2) Assess Detectability, (3) Assess Severity

Assess Likelihood

Determine the likelihood on an adverse event occurring. The risk likelihood should be determined according to a defined criteria or definition. Examples of such are shown below:

High: A standard system function or business process that has been customized by custom coding or by configuration of non-standard system parameters and/or options.

Medium: A standard system function or business process that has been significantly modified solely by configuration of standard system parameters and/or options.

Low: A standard system feature or process that has not been significantly modified by configuration or coding.

Assess Detectability

For each risk scenario a probability of detection should be determined using the following definitions or those set-down by internal company procedures.

High: errors in the output are checked by a standard system error check (i.e. integrity of data, format of data,) prior to completion of the function or process, or at the input to a downstream subsequent function.

Medium: Any errors in the output of the function will be checked by a standard system error check (i.e. integrity of data, format of data, data range) prior to completion of the function or process, or at the input to a subsequent function.

Low: Any errors in the output of the function are not checked by a standard system error check.

Assess Severity

In addition to identifying the likelihood of the risk and detectability, it is also necessary to identify effects or the severity on the business and the GxP status of the system.

High: The process is used to create, update, or process data which may have direct impact on: (1) Product efficiency (2) integrity

Medium: The process is used to create, update or process data which has direct impact upon (1) Quantity, (2) Traceability

Low: The process is used to create, update or process data that may have a direct impact on activity that supports cGxP operations.

Data Retrieval and Retention

Data integrity applies throughout the life-cycle of the data from generation through to disposal. Storage of data must facilitate retrieval when required for audit purposes or for other fact finding studies. For example batch records when generated are typically reviewed and approved. They are then archived. As time passes, archived batch records may be placed into offsite storage as their need to be onsite is not required. However, for investigations that arise or audits, the retrieval of such documents needs to be done in an orderly fashion that ensures traceability of the records while ensuring the integrity of the document is maintained. This requirement of record retrieval not only applies to paper records but to electronic records and data. Data should be "backed up" according to established procedures and must be readily retrievable when required. The practical consideration of "how" and "where" is it stored must also be carefully considered.

Failures in Data Integrity

<u>Electronic Systems</u>

The backup of test results were observed to have been imported to backup location with a different time zone designation.

User access levels did not properly a structured hierarchy of responsibilities. e.g. Operator had access to all functionality of the SCADA system.

Generic usernames and passwords were utilised to access machine recipes.

Data Integrity References and Resources

MHRA GMP Definitions and Guidance for Industry March 2015

PIC/S PI 041-1 Draft 2 (August 2016)

EMA GMP Q&A on data integrity (August 2016)

WHO Technical Report Series 996, Annex 5 (2016) Guidance on good data and record management practices

FDA Data Integrity and Compliance With CGMP Guidance for Industry (Draft) (2016)

END

www.ingramcontent.com/pod-product-compliance
Lightning Source LLC
Chambersburg PA
CBHW061218180526
45170CB00003B/1057